BELONGS TO

_____

INVENTION

INVENTION NAME : _____

THE IDEA

DATE : _____   WHERE : _____

WHY :	HOW :

THE INVENTION

FUNCTION :	USE :

PROBLEM IT SOLVES :

THE PROPERTIES

DIMENSIONS :	MATERIAL :

AUDIENCE :	PRICE :

SKETCH

NOTES

NOTES

INVENTION

INVENTION NAME : _____

THE IDEA

DATE : _____   WHERE : _____

WHY :

HOW :

THE INVENTION

FUNCTION :

USE :

PROBLEM IT SOLVES :

THE PROPERTIES

DIMENSIONS :

MATERIAL :

AUDIENCE :

PRICE :

SKETCH

NOTES

NOTES

INVENTION

INVENTION NAME : _____

THE IDEA

DATE : _____   WHERE : _____

WHY :

HOW :

THE INVENTION

FUNCTION :

USE :

PROBLEM IT SOLVES :

THE PROPERTIES

DIMENSIONS :

MATERIAL :

AUDIENCE :

PRICE :

SKETCH

NOTES

NOTES

INVENTION

INVENTION NAME : _____

THE IDEA

DATE : _____   WHERE : _____

WHY :	HOW :

THE INVENTION

FUNCTION :	USE :

PROBLEM IT SOLVES :

THE PROPERTIES

DIMENSIONS :	MATERIAL :

AUDIENCE :	PRICE :

SKETCH

NOTES

NOTES

INVENTION

INVENTION NAME : _____

THE IDEA

DATE : _____   WHERE : _____

WHY :

HOW :

THE INVENTION

FUNCTION :

USE :

PROBLEM IT SOLVES :

THE PROPERTIES

DIMENSIONS :

MATERIAL :

AUDIENCE :

PRICE :

SKETCH

NOTES

NOTES

INVENTION

INVENTION NAME : _____

THE IDEA

DATE : _____   WHERE : _____

WHY :

HOW :

THE INVENTION

FUNCTION :

USE :

PROBLEM IT SOLVES :

THE PROPERTIES

DIMENSIONS :

MATERIAL :

AUDIENCE :

PRICE :

SKETCH

NOTES

NOTES

INVENTION

INVENTION NAME : _____

THE IDEA

DATE : _____   WHERE : _____

WHY :

HOW :

THE INVENTION

FUNCTION :

USE :

PROBLEM IT SOLVES :

THE PROPERTIES

DIMENSIONS :

MATERIAL :

AUDIENCE :

PRICE :

SKETCH

NOTES

NOTES

INVENTION

INVENTION NAME : _____

THE IDEA

DATE : _____   WHERE : _____

WHY :	HOW :

THE INVENTION

FUNCTION :	USE :

PROBLEM IT SOLVES :

THE PROPERTIES

DIMENSIONS :	MATERIAL :

AUDIENCE :	PRICE :

SKETCH

NOTES

NOTES

INVENTION

INVENTION NAME : _____

THE IDEA

DATE : _____   WHERE : _____

WHY :

HOW :

THE INVENTION

FUNCTION :

USE :

PROBLEM IT SOLVES :

THE PROPERTIES

DIMENSIONS :

MATERIAL :

AUDIENCE :

PRICE :

SKETCH

NOTES

NOTES

INVENTION

INVENTION NAME : _____

THE IDEA

DATE : _____   WHERE : _____

WHY :

HOW :

THE INVENTION

FUNCTION :

USE :

PROBLEM IT SOLVES :

THE PROPERTIES

DIMENSIONS :

MATERIAL :

AUDIENCE :

PRICE :

SKETCH

NOTES

NOTES

INVENTION

INVENTION NAME : _____

THE IDEA

DATE : _____   WHERE : _____

WHY :	HOW :

THE INVENTION

FUNCTION :	USE :

PROBLEM IT SOLVES :

THE PROPERTIES

DIMENSIONS :	MATERIAL :

AUDIENCE :	PRICE :

SKETCH

NOTES

NOTES

INVENTION

INVENTION NAME : _____

THE IDEA

DATE : _____   WHERE : _____

WHY :	HOW :

THE INVENTION

FUNCTION :	USE :

PROBLEM IT SOLVES :

THE PROPERTIES

DIMENSIONS :	MATERIAL :

AUDIENCE :	PRICE :

SKETCH

NOTES

NOTES

INVENTION

INVENTION NAME : _____

THE IDEA

DATE : _____   WHERE : _____

WHY :

HOW :

THE INVENTION

FUNCTION :

USE :

PROBLEM IT SOLVES :

THE PROPERTIES

DIMENSIONS :

MATERIAL :

AUDIENCE :

PRICE :

SKETCH

NOTES

NOTES

INVENTION

INVENTION NAME : _____

THE IDEA

DATE : _____   WHERE : _____

WHY :

HOW :

THE INVENTION

FUNCTION :

USE :

PROBLEM IT SOLVES :

THE PROPERTIES

DIMENSIONS :

MATERIAL :

AUDIENCE :

PRICE :

SKETCH

NOTES

NOTES

INVENTION

INVENTION NAME : _____

THE IDEA

DATE : _____   WHERE : _____

WHY :

HOW :

THE INVENTION

FUNCTION :

USE :

PROBLEM IT SOLVES :

THE PROPERTIES

DIMENSIONS :

MATERIAL :

AUDIENCE :

PRICE :

SKETCH

NOTES

NOTES

INVENTION

INVENTION NAME : _____

THE IDEA

DATE : _____   WHERE : _____

WHY :

HOW :

THE INVENTION

FUNCTION :

USE :

PROBLEM IT SOLVES :

THE PROPERTIES

DIMENSIONS :

MATERIAL :

AUDIENCE :

PRICE :

SKETCH

NOTES

NOTES

INVENTION

INVENTION NAME : _____

THE IDEA

DATE : _____   WHERE : _____

WHY :

HOW :

THE INVENTION

FUNCTION :

USE :

PROBLEM IT SOLVES :

THE PROPERTIES

DIMENSIONS :

MATERIAL :

AUDIENCE :

PRICE :

SKETCH

NOTES

_____

_____

_____

_____

_____

_____

_____

_____

_____

_____

_____

_____

_____

_____

_____

NOTES

INVENTION

INVENTION NAME : _____

THE IDEA

DATE : _____   WHERE : _____

WHY :	HOW :

THE INVENTION

FUNCTION :	USE :

PROBLEM IT SOLVES :

THE PROPERTIES

DIMENSIONS :	MATERIAL :

AUDIENCE :	PRICE :

SKETCH

NOTES

NOTES

INVENTION

INVENTION NAME : _____

THE IDEA

DATE : _____   WHERE : _____

WHY :	HOW :

THE INVENTION

FUNCTION :	USE :

PROBLEM IT SOLVES :

THE PROPERTIES

DIMENSIONS :	MATERIAL :

AUDIENCE :	PRICE :

SKETCH

NOTES

NOTES

INVENTION

INVENTION NAME : _____

THE IDEA

DATE : _____   WHERE : _____

WHY :	HOW :

THE INVENTION

FUNCTION :	USE :

PROBLEM IT SOLVES :

THE PROPERTIES

DIMENSIONS :	MATERIAL :

AUDIENCE :	PRICE :

SKETCH

NOTES

NOTES

INVENTION

INVENTION NAME : _____

THE IDEA

DATE : _____   WHERE : _____

WHY :

HOW :

THE INVENTION

FUNCTION :

USE :

PROBLEM IT SOLVES :

THE PROPERTIES

DIMENSIONS :

MATERIAL :

AUDIENCE :

PRICE :

SKETCH

NOTES

NOTES

INVENTION

INVENTION NAME : _____

THE IDEA

DATE : _____   WHERE : _____

WHY :

HOW :

THE INVENTION

FUNCTION :

USE :

PROBLEM IT SOLVES :

THE PROPERTIES

DIMENSIONS :

MATERIAL :

AUDIENCE :

PRICE :

SKETCH

NOTES

NOTES

INVENTION

INVENTION NAME : _____

THE IDEA

DATE : _____   WHERE : _____

WHY :

HOW :

THE INVENTION

FUNCTION :

USE :

PROBLEM IT SOLVES :

THE PROPERTIES

DIMENSIONS :

MATERIAL :

AUDIENCE :

PRICE :

SKETCH

NOTES

NOTES

INVENTION

INVENTION NAME: _____

THE IDEA

DATE: _____   WHERE: _____

WHY:

HOW:

THE INVENTION

FUNCTION:

USE:

PROBLEM IT SOLVES:

THE PROPERTIES

DIMENSIONS:

MATERIAL:

AUDIENCE:

PRICE:

SKETCH

NOTES

NOTES

INVENTION

INVENTION NAME : _____

THE IDEA

DATE : _____   WHERE : _____

WHY :

HOW :

THE INVENTION

FUNCTION :

USE :

PROBLEM IT SOLVES :

THE PROPERTIES

DIMENSIONS :

MATERIAL :

AUDIENCE :

PRICE :

SKETCH

NOTES

NOTES

INVENTION

INVENTION NAME : _____

THE IDEA

DATE : _____   WHERE : _____

WHY :	HOW :

THE INVENTION

FUNCTION :	USE :

PROBLEM IT SOLVES :

THE PROPERTIES

DIMENSIONS :	MATERIAL :

AUDIENCE :	PRICE :

SKETCH

NOTES

NOTES

INVENTION

INVENTION NAME : _____

THE IDEA

DATE : _____   WHERE : _____

WHY :	HOW :

THE INVENTION

FUNCTION :	USE :

PROBLEM IT SOLVES :

THE PROPERTIES

DIMENSIONS :	MATERIAL :

AUDIENCE :	PRICE :

SKETCH

NOTES

NOTES

NOTES

Made in the USA
Middletown, DE
03 March 2022

62072270R00066